Gilbert Hilton Scribner

Where did life begin?

A brief inquiry as to the probable place of beginning and the natural

courses of migration therefrom of the flora and fauna of the earth

Gilbert Hilton Scribner

Where did life begin?
A brief inquiry as to the probable place of beginning and the natural courses of migration therefrom of the flora and fauna of the earth

ISBN/EAN: 9783337270506

Printed in Europe, USA, Canada, Australia, Japan

Cover: Foto ©berggeist007 / pixelio.de

More available books at **www.hansebooks.com**

Lieut D L Brainard
 My dear Sir.
        as promised
I send you a copy of my
little book which please accept
not for its contents or value
but as a reminder of our
delightful meeting together.
        Sincerely your friend
        J Hilton Scribner

Yonkers-on-Hudson
    21 Decr 1886.

WHERE DID LIFE BEGIN?

A BRIEF ENQUIRY AS TO THE PROBABLE PLACE OF
BEGINNING AND THE NATURAL COURSES OF
MIGRATION THEREFROM OF THE FLORA
AND FAUNA OF THE EARTH

A Monograph

BY

G. HILTON SCRIBNER

"Let the great world spin forever down the ringing grooves of change"
—LOCKSLEY HALL

NEW YORK
CHARLES SCRIBNER'S SONS
1883

TO MY LIFE-LONG FRIEND,

THE HONORABLE CHAUNCEY M. DEPEW,

WITH WHOM, IN ALL THE DEPARTMENTS OF THOUGHT, I HAVE
SPENT SO MANY AND SUCH PLEASANT HOURS, THESE
FEW PAGES ARE, WITH SINCERE FRIENDSHIP
AND EARNEST GOOD-WILL, RESPECT-
FULLY DEDICATED BY
THE AUTHOR.

PREFACE.

THE pressing duties of an ordinarily busy life, it might be supposed, would be quite sufficient to keep one not specially trained to scientific work from meddling with such matters as are presented in this brief monograph. Indeed, I have no good excuse for finding myself engaged in this business. It might, however, be said in extenuation that the topics I have been dealing with in these pages have occupied my attention from time to time, and the conclusions have been lightly held by me as mere probabilities for a long period. With later investigations and discoveries they began to crystallize into opinions, and have at length assumed the firmness of convictions. Having in an unguarded hour disclosed them to a party of friends well fitted to judge of their *possible correctness, I have by their impor tunities allowed myself to be betrayed into print.*

It has been my earnest desire, however, in writing these pages, to be as brief, con cise, and straightforward as possible in all statements of fact, even where a little more elaboration might have made a more favorable impression. I have also striven to put forth my views in a plain garb, and I shall be abundantly satisfied if I have made myself understood; more than repaid if the few and crude suggestions I have gathered shall incite abler and better equipped men to enter this very interesting field of inquiry, and bring forth such good results as I am sure await the careful and conscientious investigator; and only regretful if I have used unwit tingly any error for fact, or drawn, in the course of the argument, any false or unwar ranted conclusions.

"Inglehurst," Yonkers, Nov. 10, 1883.

WHERE DID LIFE BEGIN?

A Monograph

WHERE DID LIFE BEGIN?

THE subject of the distribution of plants and animals has for a long time engaged the attention of many able, persistent, and discriminating investigators. Much time and effort have been expended in simply observing and describing the various means by which they get about from place to place. The methods and means by which the seeds of plants are carried and deposited in new localities, the agency of insects, birds, and other animals in their distribution, no less than their own ingenious contrivances for floating with the wind and tide, and for catching on to every moving object, all have been carefully observed and faithfully chronicled.

The first important truth enforced by these observations is that all organic life on the earth is in a generic, or tribal sense at least,

migratory and nomadic.   The individuals may
be rooted and stationary, but the tribe is travelling, constantly leaving old fields and surroundings and as constantly arriving in new
ones, sometimes crowded out, sometimes
starved out, and sometimes invited out, but
always moving.   Moving on to a new environment, better suited, taking all things into
consideration, to satisfy the pressing needs of,
and to develop and raise in the scale of being,
both the individual and the species.

A second great truth taught by examining
the methods of these movements and study-
ing the causes of this ceaseless tramp of organic life is, that certain essential elements of
the environment itself are usually found to be
travelling with or a little in advance of the
migratory species.   In other words, the rainfall
and isothermal lines, the climatic and other
conditions of life, are constantly and slowly
changing relative to the locality, but moving
in fact.   It has been frequently observed that
certain species, occupying some particular territory now, have at some recent time in the
past been enabled by such changes to crowd

out other occupants of the same territory, and in turn will be undoubtedly, by similar changes and means, crowded out themselves. All kinds of plants and animals which have remained in one locality until they have lost the means of movement, which cannot or will not travel, must sooner or later first degenerate and then be exterminated. For instance, a rain-belt or an area of dew-fall veers slowly but permanently from the north to the south; an arid soil is made fertile, and a fertile soil is left arid; the grass and flowering plants in endless variety move with the dew or the rain-belt; the deer follow the grass, and the wolves follow the deer; a thousand varieties of insects follow the flowering plants, and the insectivorous birds and other animals, herbivorous and carnivorous, bring up the rear, and so on through all the interdependencies of life, the change of a single essential condition, the movement of one variety, causes a disturbance and movement of all in the neighborhood. Thence comes all this ceaseless and migratory activity among the flora and fauna of the earth.

This condition of things would indicate the possibility at least that life upon the earth had in the main commenced in some favored area, and travelled thence far and wide over the surface of the globe, driven out by changes of environment, lessening in effect the favorable conditions of its development in the place of its beginning, and ever beckoned on by more favorable conditions in adjacent districts. As there are no plants and no animals, with the exception of man, and possibly his companion the dog, and his pest the rat, that can thrive in most latitudes where any life is possible, so it is very evident that plants and animals, as we now see them, could not have made their advent upon the earth universally or simultaneously. Every geological fact contradicts both suppositions. Besides, to allege either is to claim, first, that all parts of the earth became habitable, for some form of life, at the same time, which is scarcely possible; and, secondly, such an allegation would do away with the main question of distribution, render superfluous most means of movement, and make it sheer nonsense to talk about the

time, methods, and character of the distribution of that which had from the beginning been
fully distributed. It is much more probable
that life made its first advent upon this globe
in some favored locality, and not everywhere
at once.

It would seem as axiomatic a proposition
as can be made in natural science, that life
would make its first appearance on that part
of the earth, or on that part of any developing
planet, which by climatic and all other concurrent conditions was first prepared, if not to
originate at least to receive and maintain it.
Nothing can be more certain than that it could
not make its first appearance on that part, or
on any of those parts, wanting these conditions.

By concurrent conditions of climate or temperature, wherever the phrase is used herein,
I mean such currents of air and ocean, such
evaporation and condensation of water, such
disintegration of rock, such electrical and
chemical changes, new combinations, phenom-
ena, and movements as are influenced by or
accompany changing climate or temperature,

together with all the secondary and remote
effects caused thereby. And in speaking of
the first appearance of life it matters not, to
my mind, whether it was a creation, a development, or a transplantation; whether it was a
lichen on the rock or a monad in the sea; a
single solitary primordial cell, or one molecule
of plasmic matter anywhere. This inquiry is
not for the causes, methods, character, or
extent of first life; it is simply and only con-
cerning its probable *primus locus.*

If we are so fortunate as to discover where
life began on the earth, it will be safe enough
to rest upon the assumption that much, if not
all, of the present life on the globe is its le-
gitimate result and outcome.

I.

ARE there, then, any data, any accepted facts
touching the condition of our globe antecedent to the advent of plants and animals
which would enable us to compare and con-
trast its past with its present condition, and

which under known laws would indicate what portion of the earth's surface first became, by temperature, climate, and other concurrent conditions, habitable for life? Can any reasonable, probable, and still existing cause be discovered occurring in the very centre of such first habitable portion which would have dispersed all vegetal and animal life and sent it in equal distribution through all the seas and over all the great continents as rapidly as such other portions of the earth became by temperature, climate, and other conditions ready to receive and maintain it? Is there any one locality answering to these conditions, and yet of which it may be said, in a grander and truer sense than it was said of Rome, that all roads lead to and from it; not only highways diverging to every part of the world, but with vehicles upon them; seed-wagons running constantly in the direction of the most favorable distribution and to the remotest parts of the earth? Any locality so related to the topography of the whole earth as to render such extensive movements of plants and animals from it in all conceivable directions, and to all distances, not only easy and probable, but consistent with their present distribution? Is there anything in similarity of form, anatomy, structure, size, color, food, habits, habitat, longevity, modes of propagation, terms of gestation, and capacity for inter-breeding between certain flora and fauna of the Eastern continents and the Western, which would suggest that many species and varieties so widely separated might have come originally from the same locality and ancestry? Are plants and animals always improved, developed, and rendered prolific more by being moved one way than another? Are the prevailing bottom currents of air and ocean in the direction of such favorable movements? Are cases of extermination and degeneration the result of a counter-movement, or a failure to make such favorable movements?

Many facts and considerations exist and may be presented pointing to a solution of these questions, and fairly answering some of them.

Let us consider, in the first place, the

probable condition of the earth previous to
the advent of any sort of life upon its surface. A large portion of those who have
formed any intelligent opinions, in the light
of modern thought and investigation, upon
the subject of cosmogony, believe and hold
very firmly that the earth was at one time an
intensely hot globe—indeed a molten mass,
and that in the lapse of time it has cooled
down by radiation to its present temperature.
It is not at all necessary for the purposes of
the present inquiry to examine the so-called
nebular theory, nor even to ask when or how
this globe became so heated, nor to what
extent it has now become cooled, nor need
we inquire whether the earth is now but a
molten mass covered with a comparatively
thin crust, or has cooled and hardened to its
very centre. It is important, however, to
have it understood at the outset that the
facts and considerations here presented are
addressed to those, and those only, who have
reached and adopted the conclusion that this
globe, at some time in the process of its formation and development, passed through a

fiery ordeal, that the primary rocks are of igneous formation, and that there are many other existing conditions and obvious facts which cannot well be accounted for except upon the hypothesis that the whole earth was once a molten mass.

Even after these admissions one embarrassment presents itself, happily, however, not affecting the argument, viz. :

So fully has every conceivable inference, every supposable fact and phenomenon in the development and history of the earth, been reviewed and discussed over and over again, in the light of this primitive glowing molten mass, by able and discriminating writers, that it may seem presumptuous at this late day to attempt any new deduction, or to draw any new conclusion radically important, touching this matter. But if the views here presented have been expressed before, in the relation of cause and effect, the writer has not been fortunate enough to meet with them, and it is quite safe to say that if they are correct their significance as a factor in other problems at least will not be questioned.

It is not claimed that these views have been proved to be true inductively, but there are certain facts and phenomenon pointing directly to definite conclusions hereinafter stated which I am sure every one holding and believing that the earth was at one time a molten mass will find it easier and more reasonable to admit than deny.

Regarding the earth, then, as at one time an intensely hot globe, totally destitute of organic life, one of the principal and indispensable conditions of rendering it habitable for plants and animals evidently would be the radiation into space of its excessive and destructive heat. The accomplishment of this, with the train of concurrent effects which would follow, or at least ever have followed the gradual reduction of temperature, is all that would be necessary to render the earth a suitable place for the maintenance of vegetal and animal life. At any rate this is precisely what has taken place since the commencement of the azoic age, and is still taking place on parts of the earth's surface to-day, visible and obvious to any observer.

Our inquiry, therefore, is reduced to this question : What part or parts of the earth's surface first became sufficiently cooled by radiation to be habitable by plants and animals?

A supposed case may help us in reaching a correct answer to this question. Let us assume, then, that the earth, at the time it was a molten mass, had been and was revolving in an orbit so near the sun that the amount of heat it would have been receiving from the sun would have just equalized the amount of heat it was losing by radiation. Under these conditions it would have cooled as the sun cooled—neither faster nor slower. This helps us to understand that the heat received by the earth from the sun is, and ever has been, an offset, so far as it goes, to the heat lost from the earth by radiation. A statement of the loss of heat from the earth during any definite time may be formulated in this way: From the heat lost by the earth by radiation during a given period subtract the heat received by the earth from the sun during the same period, and the remainder will be the earth's net or actual loss of heat. Sidereal heat received by the earth being infinitesimal in comparison, is not here taken into the calculation. But were it more considerable, it would not be important in this connection, for it falls upon all parts of the earth about equally.

It is evident from the present condition of the earth's surface, that at the time it was a molten mass, and for a long time thereafter, it radiated heat into space much more rapidly than it received heat from the sun; but nevertheless the heat of the sun is, and always has been, offsetting the loss of heat from the earth by radiation to the full extent of the heat which the earth had been receiving from the sun during the time.

But this sun-heat, this offset to radiation, has not been received by all parts of the earth equally. The equatorial belt, or torrid zone, has always received the most per square foot, or in proportion to its area. The two intermediate or temperate zones have received the next largest amount per square foot, or in proportion to their area; while the polar or frigid zones have received the least per square foot, or in proportion to their area.

If the amount of sun-heat received at the equator be rated at 1,000, then, upon the same basis, the average of sun-heat throughout the torrid zone should be rated at 975, the average sun-heat throughout the temperate zones at 757, and the average sun-heat throughout the frigid zones at 454, or less than one-half that of the torrid and less than two-thirds that of the temperate zones. We speak here, and shall hereafter, of the geographical zones of the earth for the sake of convenience.

The greatest amount of heat received from the sun and offsetting radiation from the earth, other things being equal, is, of course, as we have seen, at the equator, and less and less every degree north and south of this line to the poles. If, then, the frigid zones have been during all this time receiving the least heat from the sun—the least offset to their own loss of heat by radiation—does it not follow that they were the first parts of the earth sufficiently cooled to maintain vegetal and animal life?

The inference seems inevitable.

II.

BUT there are other facts leading to the same conclusion quite as suggestive in their way as those which have been cited. As every one knows, the earth is flattened at the poles and bulged at the equator, as it should be if it was once a revolving liquid globe. This gives to the polar sides an increased area of radiating surface, and to this extent an increased loss of heat. Thus it is evident that the earth is, and always has been, radiating more heat into space polewise than any other one way, and this to a limited extent has kept the polar regions in advance of the equatorial in the process of cooling.

Another effect of the same cause and bearing in the same direction is this : The equatorial diameter of the earth is about twenty-six miles longer than its diameter polewise. This condition also favors the advanced cooling of the polar regions. First, the earth is so much thinner polewise than equatorially, and consequently there is less mass per square foot to be cooled by polewise radiation than by equatorial ; or, secondly, this difference of diameters is equivalent to having had a layer around the earth of molten matter to be cooled thirteen miles in thickness at the equator, and tapering off to nothing somewhere north of the Tropic of Cancer and south of the Tropic of Capricorn in excess of the molten matter to be cooled by polewise radiation, and this would have tended to keep the polar regions constantly in advance of all other parts of the earth in cooling.

In addition to all this, it is obvious that the shape given to the earth by this difference of diameters—this flattening at the poles of itself —would somewhat lessen the angles of incident and reflection of the sun's rays within the polar regions, which would still more decrease their effect, and so reduce the offset to loss of heat by radiation within these polar regions.

Can it then be doubted that the frigid zones first became cool enough to maintain life as we now see it on the earth ?

To state the matter briefly, the polar regions have received less heat from the sun, have had less matter to cool, and have radiated heat into space more rapidly in proportion to mass than the equatorial belt or any other parts of the earth's surface. In the light of these facts it seems to me the following conclusions can-not be well avoided :

First.—That the polar zones having led the advance in cooling, have had in turn all the temperatures, climates, and climatic conditions which at any time the torrid and temperate zones have had, in addition to long later periods of cooler temperature and climates than either.

Second.—Therefore, that at one epoch or another the polar regions of the earth have enjoyed all the various temperatures and cli-matic conditions necessary to maintain all the well nigh infinite forms of life, both vegetal and animal, which are now, or ever have been, upon this globe. At the risk of being tedious let us state this hypothesis in another way.

The whole globe was once a molten mass too hot to maintain life. The polar regions were then too hot for that purpose. These same regions are now too cold to maintain

2

such life as we find on other parts of the earth. Nothing, then, can be more obvious than that the temperature of these now frigid zones, in sliding gradually from the first extreme of heat to the last extreme of cold, must have passed slowly through all the grades of temperature and climatic conditions which were exactly suited at one time or another to all the varieties of plants and animals which now live, or ever have lived on the earth.

There is no escape from this conclusion, except by asserting that the usual climatic and concurrent conditions did not in this case follow along the line of lowering temperature. But this is not only raising a groundless objection, one without a single fact to support it, but it is also one which disturbs, contradicts, and reverses the usual order of things. Certainly the *onus probandi* should rest on him who invokes the supposition.

Of course the usual number in considering this question will very properly ask whether there was time enough for organic development after the polar zones were cool enough to maintain life, and before the other parts of the earth had reached the same temperature and climatic conditions. Well, time is the infinite factor in every calculation of this sort; nature, reason, and observation unite in saying "it is illimitable and all-sufficient." Certainly there has been and will be time enough for everything.

If the first isothermal belt, including the highest heat degrees in which any life is possible had swept southward at the rate of one English mile in a millennium, it would have taken about 6,000,000 years for it to have travelled from the north pole to the equator. This would seem a sufficient lapse of time for a great advance and development of all forms of life moving within the zone. I am well aware that eminent geologists are assuming much longer periods for the life-history; for instance Professor Dana mentions 48,000,000 years as an estimated minimum of time since the commencement of the Silurian age; while Sir William Thomson estimates geological time at 100,000,000 years, Haughton at twice that period, and many others at thousands of millions. Now, without courting an ear-

nest approval for these, or any estimates for
that matter, of the lapse of time since life
began—the data being so insufficient and the
conclusions so widely different—yet it is emi-
nently conservative, in view of them and other
accepted calculations, to claim, so far as time is
a factor, that a first life-bearing climate might
have commenced anywhere and travelled everywhere (and all sorts of organisms might
have travelled with it in the natural thorough-
fares) over a globe only 25,000 miles in cir-
cumference without moving faster than one
mile in ten millenniums.

On the other hand, a zone of torrid climate
beginning near and surrounding the north
pole, and creeping thence to the equator at
such a slow pace, would have given ample
time in its long journey for the development
of highly complex forms evolved from the very
simplest, for all organisms moving within its
isothermal limits.

Considerations will be hereafter presented
showing absolutely that there was time enough
and to spare for vast and highly developed
orders of life within the frigid and temperate

zones before any movement, and probably before the equatorial belt was cool enough to maintain life, certainly before the first torrid climates near the poles became cold enough to exclude it. We may therefore safely conclude, if the code of natural laws has been uniformly in force—

First.—That life commenced on those parts of the earth which were first prepared to maintain it; at any rate, that it never could have commenced elsewhere.

Second.—As the whole earth was at one time too hot to maintain life, so those parts were probably first prepared to maintain it which cooled first.

Third.—That those parts which received the least heat from the sun, and which radiated heat most rapidly into space, in proportion to mass, and had the thinnest mass to cool, cooled first.

Fourth.—That those parts of the earth's surface, and those only, answering to these conditions, are the arctic and antarctic zones.

Fifth.—That as these zones were at one time too hot, and certain parts thereof are

now too cold, for such life as inhabits the
warmer parts of the earth, these now colder
parts, in passing from the extreme of heat to
the extreme of cold, must have passed slowly
through temperatures exactly suited to all
plants and all animals in severalty which now
live or ever lived on the earth.

Sixth.--If the concurrent conditions which
have usually followed lowering temperature
followed the climatic changes in this case, life
did commence on the earth within one or both
of certain zones surrounding the poles, and
sufficiently removed therefrom to receive the
least amount of sunlight necessary for vegetal and animal life.

It seems almost superfluous to say that
those parts of the earth which first became
cool enough to maintain life had a climate
warmer at that time than that which we now
call torrid. It was for an epoch, and probably
a very long one, as hot as it could be and
maintain life.

It is also quite obvious, in the light of the
foregoing considerations, that as the temperate
zones have always received more heat from the

sun, and have had more mass per square foot
to cool, in proportion to radiating surface, than
the polar zones, so, on the other hand, they
have always received less heat from the sun
and have had less mass to cool, in proportion
to radiating surface, than the torrid zone ; and
so when the arctic zones cooled from a tropical to what we now call a temperate climate,
the temperate zones had cooled down to that
temperature which we now call a torrid climate, while the equatorial belt was still too
hot for any form of life. Thus the lowering
of temperature, climatic change, and that life
which made its advent in these zones surrounding the poles, have crept thence slowly along,
pari passu, from these polar regions to the
equator. Doubtless, through all geologic
time, wave after wave of climatic change and
corresponding forms of life, including the remotest extinct species, from laurentian to allu-
vium, from eozoon to mammal, whose several
biographies in the rocks, are now called
epochs, have followed each other in succes-
sion from this originally prolific polar zone to
the equatorial belt.

III.

LET us now pass to a consideration of the present condition of the earth and the life upon it, and see how far things as we find them tally with the conclusions we have now reached. Before doing so, however, one or two preliminary suggestions may be advisable.

There is a very interesting, and at least plausible, theory gaining ground, that from the eccentricity of the earth's orbit, and other causes too numerous to state here, the northern and southern hemispheres are alternately submerged and drained off as the vast ice accumulations first around one pole and then around the other change slightly the centre of gravity of the earth alternately to the north and the south of the plane of the equator, and that each hemisphere makes the round of these changes in a long period or year, consisting theoretically of about 26,000, but practically, owing to the reverse movement of the precession of the equinoxes, of about 22,500 of our common years.

Whether this is so, and whether at the time of the first advent of life upon the earth, southern continents now submerged were drained off, and the northern continents submerged, or whether the northern continents were drained off and the southern submerged, as at present, is quite immaterial for the purposes of our inquiry. One thing is certain, that with the present amount of water upon the earth, to submerge the continents of the northern hemisphere as deep as the floor of the southern oceans, would necessarily drain off vast bodies of land in the southern hemisphere, and give the south pole a surrounding zone of land such as the north pole now has. Such a condition of things would keep all life in a constant migration, north and south, and almost from pole to pole—at any rate, from one frigid zone to the other.

Since then it makes no difference, for the purposes of our inquiry, whether there have been such alternate changes, and also to avoid distracting the attention by looking first to one pole and then to the other, we will consider this matter as though the continents and oceans had always been as we now find them, and confine our attention to the more fully explored arctic regions, rather than both frigid zones.

As might be readily supposed, these arctic regions which first became cool enough to maintain life, would from the same causes be the first to become too cold for the same purpose. And this cold would occur first as a temperate climate near and around the pole; at any rate, in the centre of a zone just sufficiently removed from the pole to combine the influence of the sun with its own cooling temperature, so as to become the first fit habitation of life.

This central cold creating a temperate climate would thus have become the first and all-sufficient cause of a dispersion and distribution of both the tropical plants and animals over another zone next south, next further removed from the pole, and next sufficiently cool to maintain such life. Moreover, this cooler climate occurring in the centre would have driven out and dispersed such life equally, in all possible directions. So, if the first habitable zone included the northernmost land of all the great continents which converge around the north pole, this dispersion from an increasing cold to the north of each of them would have sent southward plants and animals from a common origin and ancestry, to people and to plant all the continents of the earth, with the possible exception of Australia, whose flora and fauna are certainly anomalous and possibly indigenous.

I have mentioned cold as the all-sufficient cause of a dispersion to the southward of plants and animals. To those who would admit the cause but doubt the effect, I would quote a sentence from the admirable book of Professor Dana, entitled "The Geological Story Briefly Told," in which he says (speaking of the glacial epoch, page 224) : "There must have been some exterminations as a consequence of the cold of the glacial period and of the ice of the high latitude regions ; *many plants were driven south by the coming on of the cold, and thus escaped destruction, and some of these now live on Mount Washington and other high summits of temperate North*

America. Birds must have shortened their northward migrations and lengthened them southward, and for the most part may have escaped catastrophe—the beasts of prey, cattle, and other large mammals of Drift latitudes must also, to a great extent, have moved toward the tropics as the rigors of the approaching ice period began to be felt." If the comparatively swift coming cold of the glacial period could have moved to the southward, over vast areas, all the plants and animals of the northern hemisphere, how much more adequate and effectual for the same purpose must have been the gradual lowering of temperature during the earlier and immensely longer periods of the earth's life history.

IV.

LET us now see how admirably the earth is adapted by its surface formation and topography, for a southern migration from a zone surrounding the north pole. In the first place, nearly the whole of the earth's surface

(and all the northern hemisphere) is cor·
rugated north and south with alternate continents and deep sea-channels almost from
pole to pole. Both the eastern and western
continents extend with unbroken land con-
nections from the arctic zone through the
northern temperate, the torrid, and through
the southern temperate, almost to the antarctic
zone. Between these great continents lie the
deep oceans, whose channels run north and
south through as many degrees of latitude.
The great air and ocean currents run north
or south; all the mountain ranges of the western continent and many of the eastern continents run mainly north and south. Nearly all
the great rivers of the northern hemisphere
run north or south. To a southern migration—in other words, a migration from the
arctic region toward the equator—these pe-
culiarities of topography, these great corrugations and mountain ranges, these channels and
currents, are roads and vehicles, guides and
helps; while to an east and west migration the
same features are not only obstacles and hin-
drances, but in the main barriers insuperable.

The impassability of mountain ranges for
most plants is shown by the fact that strongly
marked varieties in great numbers and many
distinct species occur upon the eastern slopes
of the Rocky Mountains, the Sierra Nevadas,
the Alleghanies, and even lower ranges, which
are not found at all upon their western sides,
and *vice versa*.   Such a condition of things,
incompatible as it is with an eastern and
western migration, is quite consistent, however, with a north and south movement.   For
all the climatic conditions, especially that of
rain-fall, are so different on the opposite sides
of all long mountain ranges, that the same
variety split and separated by the northern
extremities of these ranges would, in moving
southward along their eastern and western
sides, and encountering such diverse conditions, have become in the course of time, under
the laws of adaptation, distinct varieties, and
probably different species.

It may be well now to examine some of the
conditions assisting this movement.   Hot air
being lighter than cold, the heated air of the
northern equatorial belt has always risen and

passed mainly toward the north pole in an
upper current, while the cooler and heavier
currents from the north have swept southward hugging the surface of the continents,
laded with pollen, minute germs and spores,
and all the winged seeds of plants, bending
grass and shrubs and trees constantly to the
southward, and so, by small yearly increments
moving the whole vegetal kingdom through
valleys and along the sides of mountain ranges,
down the great continents, always moving
with, and never across these great surface
corrugations. It is unnecessary to add that
all insects and herbivorous animals would
follow the plants, or that the birds and carnivorous animals would follow the herbivorous animals and the insects. So, too, the currents of the ocean have been established in
obedience to similar laws; as hot water is
lighter than cold, great surface currents have
been formed in both the Atlantic and Pacific
oceans, flowing from the equator to the arctic regions, while the cooler and heavier currents from the Arctic have swept the floor of
both oceans from shore to shore to the south-

ward, carrying all kinds of marine life from the pole toward the equator with them.

It may be well in this connection to allude to another fact seriously affecting the bottom currents from the pole toward the equator of both air and ocean. By reason of the revolution of the earth upon its axis, a given point upon its surface 1,000 miles south of the north pole moves to the eastward at the rate of about 260 miles an hour, while another point in the same meridian at the equator would be moving to the eastward a little more than 1,000 miles an hour; so every cubic yard of air and water which starts in a bottom current from the polar regions for the equator must, before reaching the equator, acquire an eastward motion of about 750 miles an hour. The tendency, therefore, of all bottom currents of air and ocean moving to the south, is to press to the westward every obstacle met with in its course, and the result, both as to the currents and all movable things they come in contact with, would be to give them a south-western course and movement.

Now it is a strange coincidence, if nothing

more, that the eastern coasts of all the continents have a southwestern trend, are full
of bays, inlets, and shoal water, as though
the floor of the ocean was being constantly
swept up against them; while the western
coasts are more abrupt, straight, and touch
deeper water, as though the sweepings from
the land were being constantly rolled into
the sea along their entire lines.

Notwithstanding all these indications of
a southern or southwestern movement, ever
since the migration of plants and animals first
attracted attention, students of natural science,
careful and conscientious observers, able and
discriminating investigators have, almost with
one accord, been looking east and west across
these great north and south corrugations and
natural barriers for the paths of their journey·
ings, searching along every parallel of latitude, across lofty mountain ranges, broad continents, deep and wide oceans, and ocean
currents, to and fro, and if perchance they
looked north or south it was only in search
of some ferry or ford south of the ice-fields
by which to pass the flora and fauna from one

3

continent to another, and thus account for what is very evident, viz. : That many widely distributed species and varieties have come from the same locality and had a common ancestry and origin. Is it not evident that the very plants and animals (in a tribal sense) whose migrations they have been engaged in unravelling, were as much older than ice and snow on the earth as it would require in time to lower the average temperature over a vast area from a tropical to a frigid climate ?

To give some idea of this immense lapse of time as before intimated, it may be stated that as crystallized rock is a much poorer conductor of heat than molten rock, so when the polar and temperate zones became once fairly encrusted, the main escape of heat from the earth would forever thereafter have been through the still hotter equatorial belt which was surely then receiving, as now, the intensest heat of the sun ; and so it must have remained, other things being equal, for an immense and incalculable period of time before its complete encrustation ; and even thereafter its over-heated currents of air and water

would have given to the polar regions a
torrid climate for a vast period of time.

V.

LET us now allude briefly to a few facts and
circumstances touching the existing flora and
fauna of the northern hemisphere, their past
remains and present habitat, which are entirely consistent with the views here taken,
and equally inconsistent with, and contradic-
tory of, any other cause of dispersion or course
of migration and distribution over the surface
of the earth.

The evidences of former tropical life, both
vegetal and animal, throughout the temperate zone and within the borders at least of
the arctic zone, are numerous and indisputa-
ble. It is sufficient to mention the remains,
found far within the regions of perpetual ice
and snow, of the hairy elephant, rhinoceros,
and mammoth, of the plane-tree, the palm,
and the magnolia. It is true that the ele-
phantine remains are claimed to be post-
tertiary, and they probably are; but does it

follow that either they or their ancestry abandoned a tropical climate elsewhere to invade the ice and snow in which they perished? It is much more probable that an unfavorable climate, increasing in its severity as the glacial period came on, exterminated them where their remains are found.

Elaborate theories have been contrived to account for their having been carried and deposited there, and other theories shifting the earth's axis this way and that; also asserting elevations, subsidences, and ocean currents unlike the present to furnish for them at one time or another a suitable habitat. If in the process of the earth's cooling, the polar lands and seas must have been for ages as hot as they could be and maintain tropical or any other life, it seems superfluous, to say the least (in our own time and upon the discovery of the remains of such life), to suppose subsidences and elevations to divert and extend immense warm ocean currents solely to provide for that, in later and colder periods, which was quite well enough provided for long before and in the natural order of things.

But it is a conspicuous fact that, whatever the agency evoked to keep the arctics warm, it has in each instance been assumed as a con-dition in order to account for a phenomenon, while the phenomenon itself has rested on the condition assumed. And neither the condition nor the phenomenon have at any time been necessary to account for anything if this globe has really passed through all the grades of lowering temperature from a heated mass to its present condition. It is perfectly reason-able, therefore, to suppose that these tropical forms lived when the arctic climate first be-came suitable for them and died where their remains are found; especially as nothing has yet been discovered inconsistent therewith and so much corroborative evidence points to the same conclusion.

It is a well-established fact that the Siberian elephants had an extremely long and heavy coat of hair, with an undergrowth of close, warm wool as a covering. It is scarcely less certain that they continued in the Arctics un-til the commencement of the first glacial pe-riod—the beginning of the quaternary—or the

age of man, and were then exterminated by
cold.   Can it be true that the ancestors of
these hairy elephants migrated to the Arctics?
If so, they came from a warmer climate, for
all other climates are and always have been
warmer, and they must have come from a
hairless ancestry, for all other elephants are
hairless, and so they must have come naked.
They must also have come from lands where
their food was better and more abundant than
in the Arctics, for it could not have been
poorer or scantier anywhere after lands south
of the Arctics were habitable for elephants.
Do animals having no enemies, or at any rate
no unconquerable superiors, migrate from lands
of plenty, where the climate is such that they
need no covering, to districts where food is
scarce and clothing a necessity?   It is more
reasonable, and it accords better with the law
of life, to suppose that the elephants of the
North remained in the ancient home of all
elephants, and came to have hair and wool in
their long struggle with a lowering temperature, while those that moved southward with
the southward-moving isothermal lines have

no hair because their ancestors were hairless and they have needed none. If the hairy elephant and also animals now tropical have by their remains hinted and suggested that the Arctic regions once had not only a warm climate, but actually teemed with tropical life, the remains of vegetal life have positively testified to it.

" As the tree falls, so it lies." The recently discovered coal-beds of the Arctics prove that the coal-plants once flourished there in abundance, and where the coal-plants flourished there were the warm climatic conditions of the coal-plants also. Moreover, the fossils of these coal-beds, and many of the species of plants of which they were composed, are identical with those of the coal-beds of Europe and America, showing not only a like temperature for all, but a kinship and common ancestry.

These same plants, such as tree ferns and lycopods, dwarfed and stunted by the cold of our present tropics, where only they are now found, and growing to the height of but a few feet, not exceeding a man's stature, once flour-

ished in the earlier and warmer climates of
the North, growing to the proportions of
forest trees, say from fifty to seventy-five feet.
Can it be true that the remains of these plants
found in the coal-beds of the Arctics and both
continents are telling us that the whole Northern hemisphere was at one and the same time
blessed with a uniform climate exactly suited
to them? Are they not rather *testifyiug* that
the Arctic climate was at one time suitable
for them, the climate of the Northern temperate at another time, and at a later period still
the climate of the torrid zone, in the warmer
parts of which their stunted forms still linger
on the stage? The simple fact that the carboniferous formation through these living remnants is still in sight in the tropics, and is
buried under mountains of ice and snow in the
Arctics, is to my mind evidence of many things,
and proof positive of two: first, that unless
this formation is ending where it began, which
seems almost the acme of absurdity, then it
began where it first ended, as things usually
do, and that locality is, beyond dispute, in the
Arctics; second, that as these coal-plants are

only found now in the tropics, and were once living in the Arctics, it follows that unless they came from the Arctics originally they must have travelled over and receded from a part or all of the distance and territory between the tropics and the Arctics, and in that event a warm climate, suitable for them, must have moved northward with them, or at the time of their northward movement the climate of the Arctics must have been warmer than the tropics to have invited them, which is assum· ing one useless phenomenon and two improba- bilities to account for that which is quite easy of solution in the natural order of things, and without recourse to either.

In all cases where the remains of identical forms are found in localities thus distant from each other, one of two conclusions seems in- evitable : either one came from the locality of the other, or they both had a common ances- try located in some other place. In either case the species must have travelled as far, at least, as the entire distance between the places where their remains are found ; and if they had a common ancestry located anywhere except on a straight line drawn between the two places where their remains are found, the aggregate distance of their travels must have been farther.

Apply this principle to the case of the several remains of plants of the same species found in the coal-beds of the Arctics and those of Pennsylvania, and it follows, first, that the whole distance between these widely separated coal-beds, at the least calculation, has in a tribal sense been travelled by as many species; second, that if the Pennsylvania plants did not come from the Arctics, then Arctic plants must have come from Pennsylvania, or some other locality warmer than the Arctics (for all other localities except the Antarctics are and always have been warmer), to inhabit a territory the climate of which was at a previous period precisely adapted to them, but at the time of their coming was far too cold, unless the locality they came from was as much too warm, or unless the earth was formerly heated in a method or from sources so radically different from the present that the polar regions and equatorial belt had the same climate. To state such suppositions in full furnishes for them the best possible refutation.

The most reasonable conclusion to be drawn from these facts are that the plants whose remains are found in the Arctic coal-beds lived there when the climate of the Arctics was suitable for them, and as there are no obstacles nor inconsistencies to a south-ward movement, and never have been, that all the like species found in the European and American coal-beds came generically from the same Northern locality and ancestry. But as the Arctics are now too cold for these plants, and the tropics are not, it would seem about as sound as the average geological conclu-sion that their appropriate climate came southward with them, or rather that they came with it. But, again, what is true of the course of these plants must have been true of all plants living under the same conditions, and as animals always move with and follow their food, so it is as certain that all the flora and fauna of the Northern hemisphere made a vast Southern movement from the Arctics over the Eastern and Western continents during the age of the movement of these coal-plants, as it is that the coal-plants made it themselves. And what is true of the course of migration of plants and animals of their age would, under uniform laws, causes, and conditions, be true of such movements in all ages.

VI.

A GREAT variety of opinions have been entertained and many interesting theories formed in accounting for the similarity of species and genera upon the Eastern and Western continents. If, as we have claimed, a zone around the North Pole, sufficiently removed therefrom to receive the minimum of the sun's influence consistent with animal and vegetal life, has necessarily, in its climatic progress, been in condition to maintain successively all forms of life which have ever existed on the earth, and if a region of increasing cold surrounded by this zone dispersed these various forms of life in all directions down the Asiatic, the European, and North American continents toward the equator—if this hypothesis is indeed true, and if all this occurred, we should naturally expect to find a marked resemblance in much of the flora and fauna, throughout all the continents, of the northern hemisphere. And this is precisely what we do find.

It requires but little scientific training to conclude that the mammoth, mastodon, elephant, buffalo, bison, elk, deer, hare and sheep, the wolf, fox, weasel and martin, the beaver, otter, bear, tiger, panther, lion, mountain lion and wild cat, the crocodile, alligator, frog, salmon, bass, trout and many other fresh-water fish ; that the butterflies, bees, locusts, and numerous kinds of ants and beetles ; that the endless tribes of small land birds ; that the oak, elm, maple, ash, birch, beech, larch, chestnut, and many pines, together with flowering plants, mosses, grasses, ferns and shrubs innumerable, which inhabit or have inhabited all the continents of the northern hemisphere almost indiscriminately, are related to each other, and had respectively some common ancestry and origin.

Neither the indigenous theory nor any

other hypothesis, for that matter, can or will ever account for a likeness so much more marked between certain plants and animals of the eastern and western continents than in the similarity of their surroundings, and the antecedent conditions of their habitat.

This marked similarity of the forms of life in widely separated and dissimilar environment is however possible, and only possible, in case they once migrated from the same locality, and it is abundantly safe to add that, if there is any one locality from which they all could have migrated except a northern zone, as herein described, it is yet to be discovered. Under the most plausible supposition of any other one locality, they certainly would have had to pass through some part of this northern zone to have reached their present destination. This hypothesis, of both a north and south movement in order to pass from continent to continent presents to my mind this dilemma. To suppose that plants and animals took such long and circuitous routes for the defined purpose of reaching new fields and continents, is to endow them with the forethought and intelligence of a Columbus, while to claim that they were coaxed and led along by favoring conditions is to assert that all the conditions of this long northern journey to find a suitable crossing, and afterward of an equally long southern trip to their present homes, must have been, if there has been any sort of uniformity in the thermal system of the earth of an opposite character, and therefore, if in one case they were favorable to the movement, in the other they must have been equally unfavorable ; but plants never move, and animals rarely, against conditions which are unfavorable to any extent.

It is true, so far as simple direction is concerned, that in the exceptional case of the receding of the ice at the close of the glacial period, plants and animals which had been hurried southward by the cold moved again to the northward, with the like exceptional movement of a warm climate to the northward. Both exceptions are, however, of just such a character as to prove the rule. In each case the whole movement of organic life

in the northern hemisphere was induced by a
like movement of isothermal lines and climatic conditions, and while in one instance it
was from the south to the north, the phenomenon no more suggests, when considered in
connection with its exceptional and anomalous
causes, that the general movement has not
been in the opposite direction, than the fact
that tidal waves setting up for hundreds of
miles in great rivers would indicate that their
main currents were not always and constantly
to the sea.

It hardly admits of two opinions, that organic life has in the main either moved from
the polar regions to the tropics or *vice versa*.
For, commencing in any given locality between
the two, and it could not have moved both
ways; the temperature, climate, and other
conditions north and south of every locality
are, and must ever have been, so dissimilar
that if they favored a movement in one direction they would have forbidden it in the other.
Now as east and west movements are impossible, to any great extent, and as the conditions favorable to one form of life are, as a

rule, favorable to all, it follows that the general movement of all the organic life of the earth is, and ever has been, from the warmer toward the colder, or from the colder toward the warmer, and this too from one extreme to the other, and to my mind the movement from the colder toward the warmer, from the north to the south in our hemisphere, seems not only entirely reasonable, but the facts sustaining it *positive*, and the conclusion *inevitable.*

It is a well-recognized fact and worthy of notice in this connection, that all plants and animals moved by man a few degrees from the north to the south in our hemisphere are improved and become more highly developed, vigorous and prolific by reason of the transfer, while a like movement in the opposite direction produces, in a degree proportioned to the distance moved, all the contrary effects of sterility and degeneration; so that this southern movement, in addition to all its other probabilities, is consonant with, and in the direct line of, the highest development and evolution of both plants and animals.

In the light of this fact let us ask what

changes would probably have taken place in those animals lingering behind in the arctic zone after the species to which they belonged had chiefly moved southward, remaining there until its tropical climate had become temperate and then frigid. These remnants, in their struggle for adaptation to the new conditions of increasing cold would, after passing through the temperate and encountering the frigid climate, doubtless have been exterminated or would have become degenerations, like the whale, the walrus, the sea-lion, and the whole seal tribe of the present arctic regions, receding slowly toward the water and cold-blooded life from which possibly all animal life originally came.

These degenerations themselves furnish some proof that the arctic regions once had a warm climate; to hold otherwise is to allege that their ancestry forsook a favorable climate for one in which they could only escape extermination by degenerating and taking to the water for subsistence. I think it more reasonable to assume that the favorable climate forsook them, and once caught in unfavorable conditions they have been going on from bad to worse in a fierce struggle for existence.

Evolution and degeneration in the organic world are, in one phase at least, the result of changes in the counter relations of demand and supply. Whenever in respect to the organism, and all things possibly useful to it, the supply by the smallest degree exceeds the demand, old wants and capacities are enlarged, and new wants creep into existence; therewith old organs are improved and new ones developed for securing and appropriating by defence, contest, and competition, this surplus; the organism thus acquires varied appetites, increased activity, diversified employments, keener sensibilities, and a wider range of life, and so passes by such changes from the simpler to the more complex in form and function—and this is evolution. So, on the other hand, when the demand is greater than any available supply, the wants denied must be suppressed; therewith the organs and capacities specialized for their gratification first fall into disuse, are then atrophied, and thus the organism with restricted activity, limited employments, and a narrower range of life, recedes from the complex to the simpler in form and function—and this is called degeneration. It is all the result of the changing relations of demand to supply. And in the very last analysis, in every case, I more than suspect that all life which occurs within a certain range of heat is the demand, and that all heat which is suitable to such life-range is the supply.

These cases of degeneration in different orders are more numerous than was formally supposed. They have, in fact, been just as frequent as the permanent success of a species (by limiting their needs) in their struggle for adaptation to an adverse environment, one which continually diminished the variety and quantity of supplies, and yet changed in its unfavorable conditions so slowly as not to exterminate the species.

This fact offers us another suggestion in this connection, If it is true that, in common with many existing plants and animals, the ancestry of man—some animal with a thumb, and so having the possibility of all things—shared this northern home, this common and immensely remote origin, earlier by long epochs than the glacial period, it would afford a possible ground for the claim of the unity of the origin of man, and also a reason for the absence on the earth of his immediate predecessor. His arboreal progenitor in the pioneer ranks of this great southern movement, ages before the quatenary (during all of which period man has probably inhabited the earth), was possibly driven naked by the ever following, merciless cold, thus keeping him within the southward moving tropical climate, down the eastern and western continents alike, until it and he, arriving in the lapse of ages at the equatorial belt, and being always at the head and still rising in the scale of being by this movement, discipline, and process, became sufficiently advanced by slow degrees to build fires, clothe himself, make implements, and, possibly, domesticate animals, at least the first and most useful to primitive man, the dog, and so prepared for conflict and for all climates, turned backward to the verge of everlasting ice, subduing, slaying, and exterminating, first his own ancestry, his nearest but now weak rival, which by lingering behind and struggling for life in a climate of increasing cold, would have become extremely degenerated and so easily disposed of, if not actually exterminated, by the climate itself, thus leaving as the nearest in resemblance to man, and yet the remotest in actual relationship both to him and his ancestry, the later tribes of anthropoid apes since developed nearer to the equator, from the next lower animals which accompanied him in his southward march.

This last proposition, however, is but a vague and very deductive supposition, for which nothing is claimed beyond a possibility, or bare probability. Notwithstanding, to sustain the main conclusion herein stated, with all the essential results and outcome of the same, it seems to me to be only necessary to claim, what would be generally admitted, viz. : that the whole earth was at one time too warm to maintain life, and that no discovered cause or fact points to a less average difference between

the temperature of the tropics and the polar
zones in the past than we find existing between
them to-day. For even such a condition and
difference of temperature would have given
to the polar regions first a tropical climate
and life, then a temperate climate and life
appropriate to it, and each for an immense
epoch before the equatorial belt would have
been habitable for any known organism.

VII.

ONE glance at the immediate cause—the
proximate moving power—of all this vast, varied, and complex machinery and movement
of life, and I have done. We have been thus
far contemplating and discussing the results
of this power ; attempting a partial description
of the methods made manifest in its grand life-
bearing and eliminating work ; discovering its
ways in prompting, developing, sifting, and
destroying life ; following in the paths and examining the effects of its great zones or belts
of graded climates girding the earth, and which

have swept from the poles to the equator—one after the other, and age after age, through a period so vast, so incomprehensible, that even the epochs of which it is composed seem to us almost like so many eternities. Let us draw near, observe and define more closely this wonderful force, an excess of which forbids life, a deficiency of which destroys life, and without which life is impossible.

Heat is the proximate cause of all activity. With it life comes and goes. The departure of heat which we call "cold" is death. Cold huddles and quiets the molecules of all known substances. Life cannot invade its precincts. As cold dispersed the plants and animals from the polar regions, so it has set up an impassable barrier against their return. Every high mountain has a cold line above which nothing thrives. The earth is growing colder age by age. Our next neighbor, the moon, has already become cold and lifeless, while yet receiving its full proportion of sun-heat. Certain grades or degrees of heat then not only constitute the moving power of all life, but does it not seem probable that the constant lowering temperature of the earth, the constant loss of heat during its life-history, with the concurrent effects thereof, has been directly or indirectly the great and all-sufficient exterminator of extinct species?

Of course heat and cold are comparative terms. But notwithstanding, as the whole globe was once too hot, and certain portions of it are now too cold for life, it follows conclusively that there is a definite range of temperature, a fixed number of degrees of heat, which constitutes the gamut of life. No organism known can exist an instant above or below it. The numerous subdivisions of this life-scale, both cause and define the con-ditions most favorable to the development of the several species and varieties of plants and animals. But a portion of these subdivisions, viz. : the higher heat lines of this great life-range have passed from the earth forever, or rather, the earth has passed through them, dropping from time to time the extinct species which were only fitted for hotter climates, while the cooler subdivisions, the lower heat lines within this gamut of life, we are now passing through, and they now constitute, as their predecessors once did, the great zones of temperature—the isothermal lines encircling the globe and moving forever slowly from the poles to the equator, each bearing with it, developing, and raising in the scale of being its peculiar forms of life.

Earth's wrinkled crust reveals to us the beginnings of life, and our own age gives plain indications of its ending. The laurentian rocks stood god-father to the first-born, and to-day the death-line encircling the poles, drawn where life first began, studded with white pinnacled monuments, guards from intrusion the cemetery of departed ages. The last life on earth may be as remote in the dim future as the first is in the shadowy past, but the indications are that within the polar regions we have now the beginning of the end.

Thus the arctic zone, which was earliest in cooling down to the first and highest heat degree in the great life-gamut, was also first to become fertile, first to bear life, and first to send forth her progeny over the earth. So too, in obedience to the universal order of things, she was first to reach maturity, first to pass all the subdivisions of life-bearing climate, and finally the lowest heat degree in the great life-range, and so the first to reach sterility, old age, degeneration, and death. And now cold and lifeless,—-wrapped in her snowy winding sheet, the once fair mother of us all, rests in the frozen embrace of an ice-bound and everlasting sepulchre.

APPENDIX.

"In scientific investigations it is permitted to invent any hypothesis, and if it explains various large and independent classes of facts, it rises to the rank of a well-grounded theory."—CHARLES DARWIN : *Animals and Plants under Domestication*, vol. i., page 9.

"The globe, when its continental area had become in the main *terra firma*, may hence have had other great areas unsolidified."—DANA's *Manual of Geology*.

"That tendency which we see in the human races, to overrun and occupy each other's lands as well as the lands inhabited by inferior creatures, is a tendency exhibited by all classes of organisms, in all varieties of ways."—HERBERT SPENCER : *Principles of Biology*, vol. vii., page 315.

"Any alteration in the temperature of a climate or its degree of humidity, is unlikely to affect simultaneously the whole area occupied by a species; and further, it can scarcely fail to happen that the addition or subtraction of heat or moisture will give to a part of some adjacent area a climate like to that to which the species has been habituated."—HERBERT SPENCER: *Principles of Biology,* vol. i., page 428.

"Since in other periods we know that life was always present when its conditions were present, it is not unreasonable to look for the first traces of life in this formation (the lower Laurentian), in which we find for the first time the completion of those arrangements which make life, in such forms of it as exist on our planet, possible."—DR. J. W. DAWSON : *Retiring address as President of the American Association for the Advancement of Science,* 1883.

"As so many animals are dependent on vegetation, its changes immediately affect their distribution."— ALFRED RUSSEL WALLACE : *Distribution of Animals,* vol. i., page 43.

"There are few, I presume, who reflect on the subject that will not readily admit that, whether as regards the great physical changes which are taking place on the surface of the globe, or as regards the growth and distribution of plant and animal life, the ordinary climatic agents are the real agents at work, and that compared with them all other agencies sink into insignificance."—CROLL'S *Climate and Time.*

"Even in the arctic zone there were in the miocene great forests of beach, oak, poplar, walnut, and red-

wood, with magnolias, alders, and others."—JAMES D.
DANA, LL.D. : *The Geological Story briefly Told*, page
200.

"At the same time (the miocene), or perhaps somewhat earlier, a temperate climate extended into the
arctic regions, and allowed a magnificent vegetation
of shrubs and forest trees, some of them evergreen,
to flourish within twelve degrees of the Pole."—AL-
FRED RUSSEL WALLACE : *Distribution of Animals*, vol. i.,
page 41.

"Coal-beds of carboniferous age are extensively
developed in the arctic regions."—JAMES CROLL : *Climate and Time*, page 198.

"The woolly rhinoceros, on the other hand, may be
viewed as a northern form, since it is met with in
vast abundance in the arctic regions of Siberia as well
as in Europe, and has not been found south of the
Alps and Pyrenees."—W. BOYD DAWKINS, M.A.,
F.R.S., F.G.S. : *Cave Hunting*, page 400.

"That an equable condition of climate extended
to near the North Pole is proved by the fact that in
the arctic regions vast masses of carboniferous lime-
stone, having all the character of the mountain lime-
stone of England, have been found."—JAMES CROLL :
Climate and Time, page 297.

"This writing and these figures consist of the remains of animals and plants which, in the great majority of cases, have lived and died in the very spot in which we now find them, or at least in the immediate vicinity."—T. H. HUXLEY : *Origin of Species*, page 42.

"It follows . . . that man, issuing from a 'mother-region' still undetermined, but which a number of considerations indicate to have been in the north, has radiated in several directions ; that his migrations have been constantly from north to south." —M. LE MARQUIS G. DE SAPORTA : *Popular Science Monthly*, October, 1883, page 753.